I0469322

Fire at Watts Bar
Hydroelectric Plant

Authored by: Jennifer L. Roberson
Hollis Stambaugh

This is Report 147 of the Major Fires Investigation Project conducted by Varley-Campbell and Associates, Inc./TriData Corporation under contract EMW-97-C0-0506 to the United States Fire Administration, Federal Emergency Management Agency.

Homeland
Security

Department of Homeland Security
United States Fire Administration
National Fire Data Center

U.S. Fire Administration Fire Investigations Program

The U.S. Fire Administration develops reports on selected major fires throughout the country. The fires usually involve multiple deaths or a large loss of property. But the primary criterion for deciding to do a report is whether it will result in significant "lessons learned." In some cases these lessons bring to light new knowledge about fire--the effect of building construction or contents, human behavior in fire, etc. In other cases, the lessons are not new but are serious enough to highlight once again, with yet another fire tragedy report. In some cases, special reports are developed to discuss events, drills, or new technologies which are of interest to the fire service.

The reports are sent to fire magazines and are distributed at National and Regional fire meetings. The International Association of Fire Chiefs assists the USFA in disseminating the findings throughout the fire service. On a continuing basis the reports are available on request from the USFA; announcements of their availability are published widely in fire journals and newsletters.

This body of work provides detailed information on the nature of the fire problem for policymakers who must decide on allocations of resources between fire and other pressing problems, and within the fire service to improve codes and code enforcement, training, public fire education, building technology, and other related areas.

The Fire Administration, which has no regulatory authority, sends an experienced fire investigator into a community after a major incident only after having conferred with the local fire authorities to insure that the assistance and presence of the USFA would be supportive and would in no way interfere with any review of the incident they are themselves conducting. The intent is not to arrive during the event or even immediately after, but rather after the dust settles, so that a complete and objective review of all the important aspects of the incident can be made. Local authorities review the USFA's report while it is in draft. The USFA investigator or team is available to local authorities should they wish to request technical assistance for their own investigation.

For additional copies of this report write to the U.S. Fire Administration, 16825 South Seton Avenue, Emmitsburg, Maryland 21727. The report is available on the Administration's Web site at http://www.usfa.dhs.gov/

U.S. Fire Administration

Mission Statement

As an entity of the Department of Homeland Security, the mission of the USFA is to reduce life and economic losses due to fire and related emergencies, through leadership, advocacy, coordination, and support. We serve the Nation independently, in coordination with other Federal agencies, and in partnership with fire protection and emergency service communities. With a commitment to excellence, we provide public education, training, technology, and data initiatives.

ACKNOWLEDGMENTS

The U.S. Fire Administration greatly appreciates the cooperation received from the following people and organizations during the preparation of this report:

U.S. Tennessee Valley Authority (TVA)
David J. Icove, P.E., TVA Police
Thomas K. Heffernan, Safety, Transmission Power Supply, TVA
H. Lee Hustead, Training Supervisor, TVA

Rhea County Emergency Management
Billy Cranfield, Rhea County Emergency Manager

Wolf Creek Fire Department

Spring City Fire Department

Evensville Fire Department

TABLE OF CONTENTS

EXECUTIVE SUMMARY 1

SUMMARY OF KEY ISSUES 2

BUILDING STRUCTURE AND SITE 2

INCIDENT NARRATIVE 6

    Injuries and Fatalities 9

THE INVESTIGATION 10

    Damage Assessment 11

KEY ISSUES 12

    Building Structure 12

    Fire Protection Systems 13

    Communications 13

    Risk 15

    Pre-planning 15

    Training 16

    Power Generation and Terminal 16

    Compliance with Life Safety Code 17

FIRE ORIGIN AND SPREAD 17

LESSONS LEARNED 18

Fire at Watts Bar
Hydroelectric Plant
September 27, 2002

EXECUTIVE SUMMARY

On September 27, 2002, at approximately 8:30 a.m., the Rhea County, Tennessee 9-1-1 Center began receiving calls about a fire at the Watts Bar Hydroelectric Plant. The plant, built between 1939-42 and operated by the U.S. Tennessee Valley Authority (TVA), is constructed of steel and concrete, and sits on the Tennessee River midway between Knoxville and Chattanooga. The dam supplying water to the hydroelectric plant is 112 feet high, and approximately one-half mile long. The plant has a generating capacity of 175,000 kilowatts. It supplies power for TVA, and provides back-up power for the Watts Bar Nuclear Power Station located directly south of dam.

At the time of ignition—estimated to be 8:15 a.m.—there were five employees working in the hydroelectric plant control room. The fire spread rapidly, giving these personnel only four minutes to realize that there was a fire and to escape. All five were able to evacuate the control building, although each employee suffered smoke inhalation. There were no injuries.

The first call concerning the fire was made by a water delivery man who was at the control building. A second call was made from the Watts Bar Lake Resort located down the street from the plant.

The fire began in the vertical cable shaft and spread so rapidly to the control building and burned so intensely that fire suppression personnel were unable to make entry into the building until 9:10 a.m. The fire self-extinguished due to lack of fuel at some point, although investigators could not determine precisely when this occurred. The intensity of the heat generated during the fire prevented firefighters from accessing the seat of the fire until 12:17 p.m. In all, over forty personnel from both volunteer and industrial fire departments responded to the fire.

A multi-agency investigation team lead by the TVA Police began the origin and cause investigation on the following day, September 28th. Based upon physical evidence, the area of origin was determined to be within a 120-foot vertical cable shaft running from the hydroelectric plant to the control building. Fire modeling was then used to determine fire development and spread, and to estimate the temperature of the fire. The hydroelectric plant has remained closed since the fire, at a business loss of approximately $100,000 per day.

Although many people perceive a steel and concrete structure as not being a fire risk, this fire illustrated, once again, that a fire in an electrical system can quickly be life-threatening regardless of the structure type.

SUMMARY OF KEY ISSUES

Issue	Comments
Building Structure	The concrete and steel construction of the building trapped heat, elevating temperatures. In addition, the plant had a constant breezes flowing through it, from the dam up through the control building. Five people nearly became trapped due to their location in the building, rapidity of fire development, and difficulty of egress.
Fire Protection Systems	There were no fire suppression systems within the control building. The annunciator alarm wires were severed as a result of the fire.
Communications	Due to topography, there was very little radio reception. Moreover, radios for the Watts Bar Nuclear Power Station fire brigade must maintain a weak signal due to the power station. Most decisions and communications were face-to-face. Once the cables in the shaft were cut, communication with personnel in the powerhouse was lost. Employees there were unaware of fire in the control building.
Risk	The potential for a fatality was great, both within the control building and the powerhouse. Injuries to personnel in the powerhouse were prevented because a senior supervisor had a flashlight and led the employees to safety.
Pre-planning	Local responders had no pre-plans, and had never been through the building except as tourists. Emergency plans for the building were kept in the building, and as a result were inaccessible to responders.
Training for Response Personnel	Local responders and WBN fire brigade personnel were unfamiliar with tactics for fighting a fire in this type of structure. Additionally, firefighters were unfamiliar with TVA tactics for fighting electrical fires.
Power Generation and Transmission	There was uncertainty about whether some areas of the building were still electrically charged. In fact, while power to the structure had been cut, the batteries retained their charge for another full day.
Compliance with Life Safety Code	WBH had been grandfathered in under the Life Safety Code, and so was non-compliant. This is now being addressed by TVA.

BUILDING STRUCTURE AND SITE

The Watts Bar Hydroelectric Plant is located at mile marker 529.9 of the Tennessee River, sitting halfway between Knoxville and Chattanooga, in Rhea County. It is one of nine TVA dams located on the river, and has been in continuous service since it opened in 1942. Originally constructed to provide power to Oak Ridge Laboratories and the public, and flood control for the Tennessee Valley, the plant still provides those services along with backup power to the Watts Bar Nuclear Power Station and cooling water for the reactor. The plant is bordered to the north by State Road 68, to the West and South by the Watts Bar Nuclear Power Station complex, and to the East by the Tennessee River and Miegs County. The hydroelectric plant consists of a powerhouse, a control building, and a switchyard. The hydroelectric plant sits adjacent to the Watts Bar Nuclear Power Station, the two facilities are operated by different TVA department.

The hydroelectric powerhouse sits directly on the Tennessee River, and consists of five 30,000-kilowatt unites. Water is fed directly into turbines to generate power, while excess water flows through sluice gates connected to the powerhouse. Power created by these units is fed from the powerhouse to the control building and the switchyard via copper cables covered in butyl rubber insulation. The cables run directly from the generators to the switchyard by way of an L-shaped cable shaft. The shaft runs horizontally west from the generators for approximately 200', then runs vertically from the powerhouse to the control building and cable switchyard through a 120' shaft. The tunnels are constructed of concrete, with cable trays lining the walls on both sides of the horizontal shaft. The five generator leads are carried through a concrete conduit located on the floor in the middle of the horizontal shaft The diagram to the right shows a cross-section of the horizontal shaft.

Ten platforms are located in the vertical cable shaft. The platforms are made of steel grating, and are spaced vertically at ten-foot intervals. The platforms can be used to visually inspect the cables, and to perform maintenance. Platforms are accessed and connected by steel ladders—one ladder connects two platforms, with ladders on alternating sides of the shaft as they ascend.

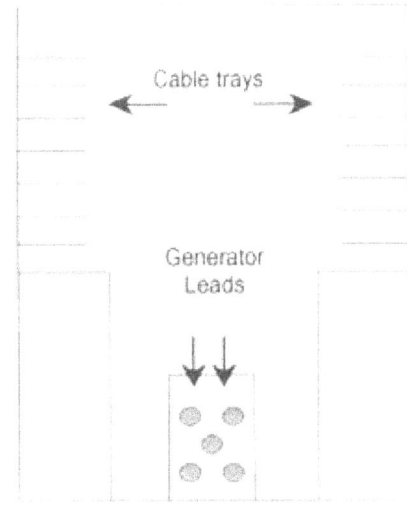

Cross-section of the horizontal cable shaft

More specifically, the cable shaft contained several types of electrical cable including:

- Generator leads for five hydroelectric units consisting of two 1750 MCM (Thousand Circular Mils) diameter cable per phase running in conduit and a concrete duct bank at the west end of the shaft.

- One set of 480V cable running from two separate 480V Main Auxiliary buses with 1000 MCM cable per phase running up both the north and south sides of the shaft;

- Two sets 250V DC "00" cable running up both the north and south sides of the shaft;

- Over 150 multi-conductor instrument and control cable at the east end of the shaft;

- Telephone cable; and

- Annunciator fiber optic cables.

Prior to the fire, the cables running through the shaft were coated in a variety of substances, including the previously mentioned butyl rubber and polyvinyl chloride (PVC). The 480V cables were mounted in the shaft using wood blocks to prevent their contact with the metal platforms. Running parallel to the cable shaft is a visitor's elevator. Before the fire, the elevator was used to show visitors the powerhouse as a part of their tour of the plant. At the time of the fire, there were no visitors at the plant.

The control building is located on the south side of State Road 68 on a cliff above the powerhouse, and consists of a visitor's center, control room, terminal room, cable spreading room, and other smaller areas. The building consists of two main levels: the control room level and the terminal room level. Prior to the fire, the control room level was accessible via a pedestrian overpass from the north parking lot, and consisted of several large windows overlooking the Tennessee River and the powerhouse. The terminal level was accessible via stairs from the control room level, or through a door from the south western side of the building. Entrance from that point was offset −people entering through the southwest doors would find themselves on a landing between the two floors, with a half flight of stairs leading up or down. The terminal room level also had several small windows, located along the top of the east side of the building and directly next to the southwestern door.

View of the Control Building from
the Dam

View of the Control Building from across Route 68

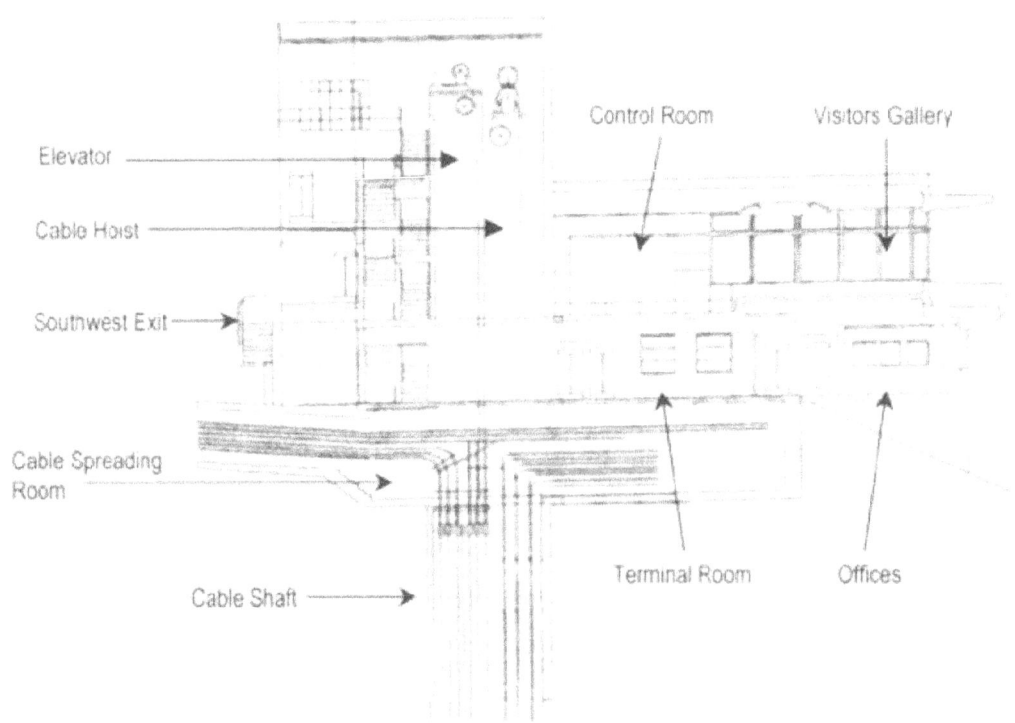

Watts Bar Hydroelectric Plant Control Building

The tops of the elevator and cable shafts were also located within the control building. The tops of both shafts contained machinery not only for the operation of the elevator, but also to hoist or lower cables and steel platforms. Below the terminal level was the cable spreading room. It was here that the cables terminated their vertical ascension and once again, began to run horizontally. Contained in this room was a small flight of stairs, which could be used by an engineer or other employee to inspect the cables before they ran underground into the cable switchyard.

Due to the hydraulic properties of the site, a constant breeze flows from the powerhouse floor, through the cable and elevator shafts, to the control building At a minimum, this breeze is 1/8 miles per hour—any fluctuation in wind speed outside the structure increases the velocity of the breeze. The chimney effect resulting from this natural phenomenon contributed significantly to the fire development and spread. On the day of the fire, the wind speed outside the plant was 16 mph.

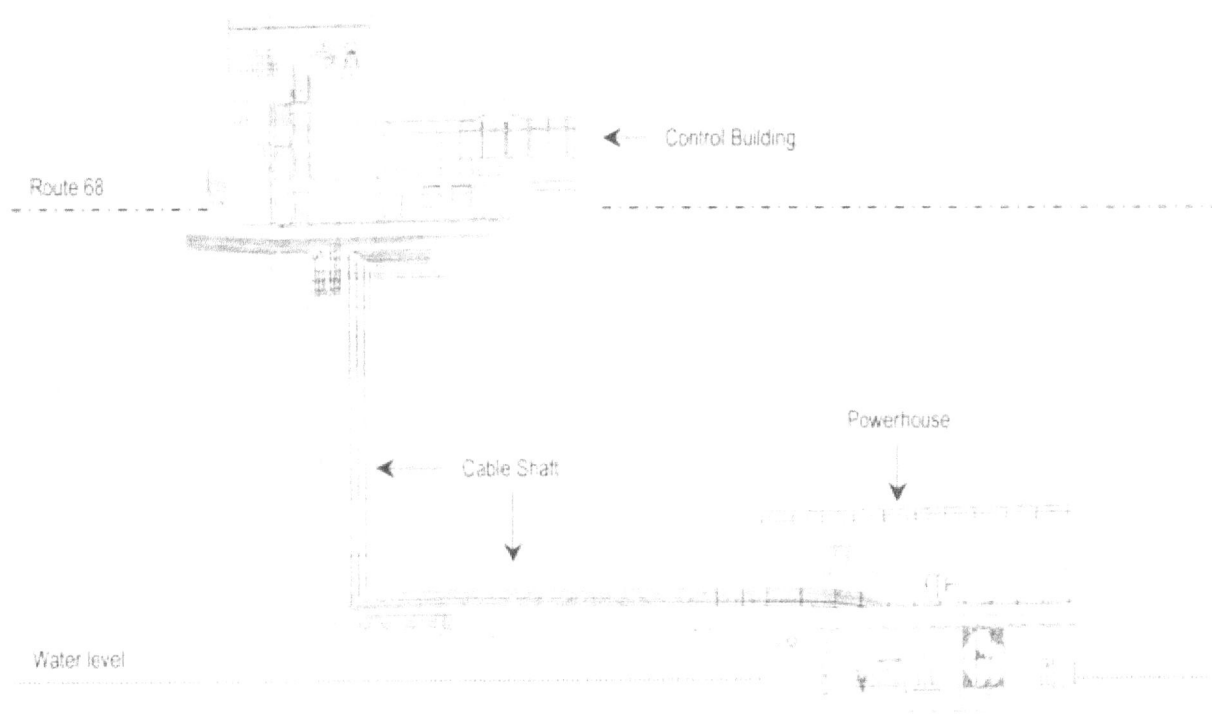

Side view of the Watts Bar Hydroelectric Plant

INCIDENT NARRATIVE

At approximately 8:15 a.m. September 27, 2002, a phase-ground-phase fault occurred from two 480 Volt AC power cables and a steel grating located in a vertical concrete cable shaft. The control cables carried power from the powerhouse to the control building for normal operation of the plant and the plant equipment As previously described the control cables (located at water level), traveled up to a cable separating room and the switchyard by means of a 120-foot vertical shaft. The electrical fire started in the vertical shaft and rapidly became a ventilation-based fire, which was fed by the large number of cables and the constant breeze running through the shaft. According to the National Weather Service, weather conditions on the day of the fire were; temperature of 72 degrees F, humidity 83%, and a wind speed of 16 mph.

At 8:24 a.m., the Watts Bar Hydroelectric Plant Line "A" tripped, indicating that something was wrong within the plant. At approximately the same time, five workers who were located directly above the cable shaft noticed smoke. In the four minutes it took workers to evacuate, fire conditions worsened, and smoke spread throughout the building. The workers successfully exited the building despite suffering from the effects of smoke inhalation.

Since the fire intensified so quickly, and it was imperative to evacuate immediately, workers were not able to all 9-1-1 or the plant prior to fleeing. The first 9-1-1 call was made by a water delivery-man who notified his dispatcher of the fire, and

Control building and switchyard, looking southeast on Highway 68

requested that they call 9 1-1. A second call was made to the Rhea County 9-1-1 center by a Watts Bar employee. Rhea County began receiving these calls at 8:30 a.m. The Rhea County Emergency Manager was notified of the fire at 8:31 a.m. Watts Bar Nuclear Power Station, located directly south of the Hydroelectric Plant, dispatched members of their Fire Protection Unit to assist with fire suppression activities at approximately the same time.

At 8:35 a.m., units from Wolf Creek Fire Department and Spring City Fire Department arrived on the scene at the southwestern side of the control building. Simultaneously, the fire brigade from TVA's Watts Bar Nuclear Power Station arrived at the north parking lot with their Tele-squirt. The fire brigade immediately requested additional off-site assistance, with the result that units from Pine Grove Fire and Evensville Fire Department responded to the scene.

The units on-scene were informed that the hydrants located in the plant switchyard were inoperable, and that water supply would be drawn from hydrants located directly off State Road 68. 500' of 5" supply line was laid from the Tele-squire in the north parking lot to a hydrant on the State road. The Tele-squirt then supplied a 3" line laid from the apparatus to the north end of the pedestrian footbridge. Running off the 3" line were two 1 ¾" attack lines. In addition, units from Pine Grove Fire Department had laid 900' of 5" supply line from a hydrant on the northwest side of the switchyard to one of their engines on the south side of the building. This engine supplied a 2 ½" line to a wye, which then fed two uncharged 1-1/2" attack lines. Water supply was successfully established at 9:10 a.m.

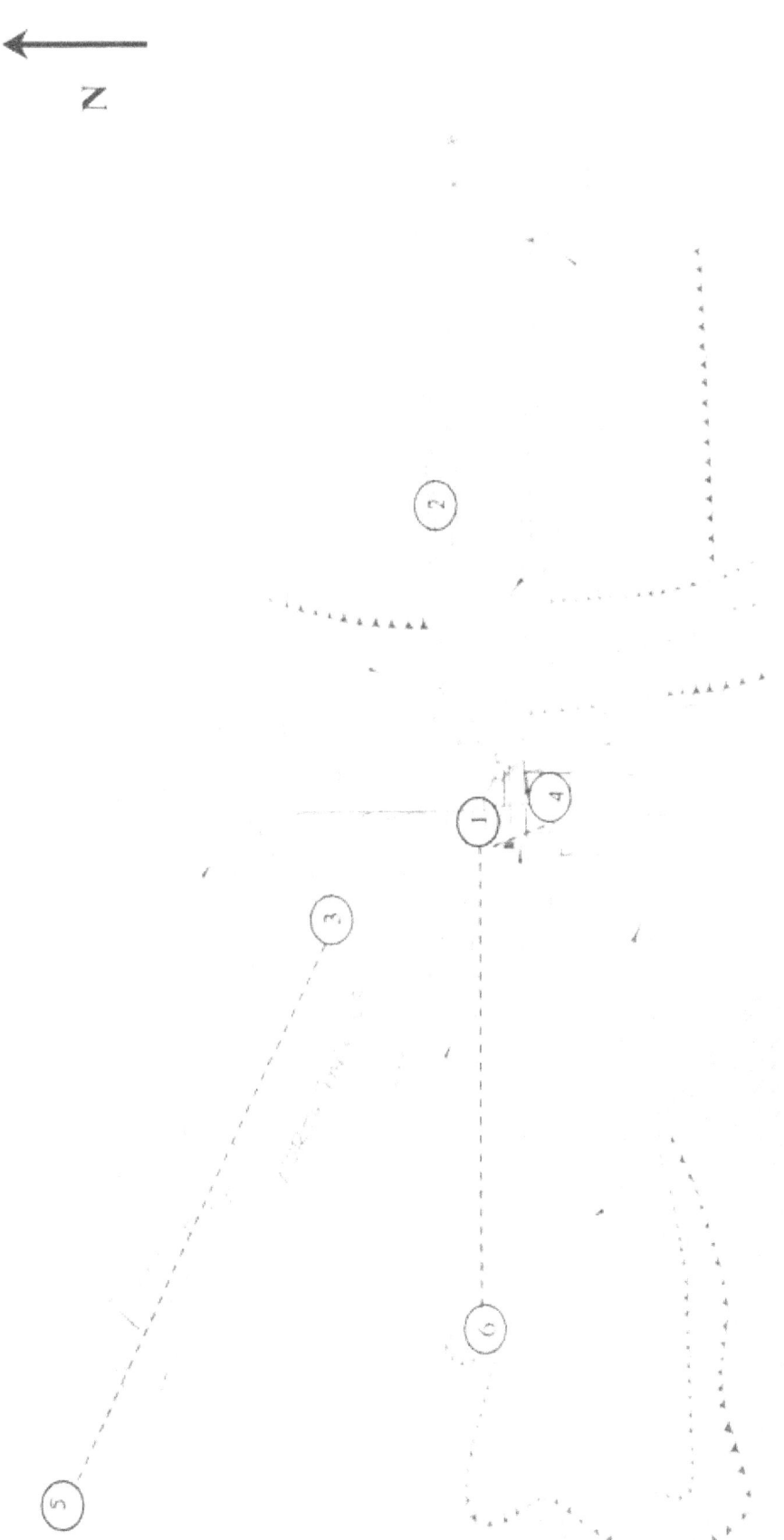

Site Diagram and Apparatus Location

1. Control Building
2. Powerhouse
3. Location of Watts Bar Nuclear Power Station Tele-Squirt
4. Location of volunteer units
5. Location of Hydrant 1
6. Location of Hydrant 2

Initial entry into the control building was made at 9:11 a.m Attack crews crossed the pedestrian footbridge on the north side of the building, and made entry into the control building via the control room. Fire personnel used a thermal imaging camera to locate the seat of the fire, and attacked the fire using a 1-3/4" attack line Due to the severity of the heat, however, firefighters backed out of the building at 9:20 a.m. and an alternative attack strategy was developed. A second attack was attempted upon the order of the Watts Bar Nuclear Power Station unit commander who at the time was functioning as the Incident Commander. At 9:35 a.m., the two 1-1/2" attack lines running from the Pine Grove engine were charged and advanced on

Ventilation attempt on the southwest side of the building

the southwest door. Firefighters forced entry through the door at 10 a.m and proceeded to knock down a small amount of fire before backing out due to the heat. After this attack, firefighters from Watts Bar Nuclear Power Station hydraulically ventilated the visitor's center/control room using a 1-3/4" hand line, and then a smoke ejector. During ventilation, three firefighting teams of two personnel each rotated the fire attack through the southwest door, with each team making three attacks and further knocking down the fire.

Command personnel meeting to discuss strategy

Personnel from the TVA Sequoyah Nuclear Power Station (SQN), located outside of Chattanooga, arrived on-scene at 10:47 a.m. reported to the unified command post in the north parking lot, and immediately began assisting with fire suppression. Since Federal regulations governing fire protection of nuclear power plants limit the amount of time that scheduled fire personnel can be off-site, firefighters from the Watts Nuclear Power Station Fire Protection Unit were required to return to their posts within two hours. Likewise, at 11:45 a.m., personnel from the Watts Bar Nuclear Power Station Fire Brigade returned to the Nuclear Power Station. At this time, incident command was transferred from the Watts Bar Nuclear Power Station Fire Brigade to the Rhea County Emergency Manager.

At 11:50 a.m., personnel made entry to the Terminal Room, believed at that point to be the seat of the fire. Attack crews extinguished three fires in the cables and terminal boards, and exited at 12:05 p.m. Even with continuing ventilation, heat in the room remained overwhelming. In order to improve ventilation, the west exterior windows of the building were removed. Crews then re-entered the control room, extinguished a small fire in the relay board, and vented into the visitor gallery by breaking a large window separating the two rooms. Crews exited the building at 12:30 p.m.

To release the heat from below the control building, crews re-entered the building at 12:50 p.m. to open the hatch to the cable shaft. The crew exited the building after successfully completing their mission at 1:09 p.m. Further reconnaissance of the building was required to determine fire spread. At 1:20 p.m., an attack crew was directed to enter the building advance to the cable spreading room, and if possible, check the battery room. The crew was able to advance to the cable spreading room where they found a small fire on a cable reel, smoldering fires in filing cabinets, and almost complete burnout. The attack line was not long enough to reach the room so firefighters used hose from the damaged hose rack to extinguish the cable reel fire. Once the fire was extinguished, firefighters checked the lower battery room, but were unable to determine the extent of damage. They then checked the cable tunnel and found severe fire damage in the first fifteen feet. Firefighters were forced to leave the building at 1:40 p m. when their SCBA alarms activated.

Briefing prior to first entry to cable gallery

Burned cables, running from the cable tunnel to the switchyard

At 1:42, the TVA Public Information Officer was informed that the fire was extinguished and overhaul was in progress. Unit commanders met to discuss their course of action, while personnel checke the Con rol Room le el for further fire and performed additional ventilation. A crew checked the terminal room level for fire at 2:15 p.m., and found only small spot fires burning. At 2:25 p.m., crews began systematic entries into the building to perform overhaul and cool the structure as much as possible. At 3:15 p.m., crews again checked all floors of the building and the cable tunnel to see that there were no further fires and that the areas were cooling. An additional check for spot fires in the cable spreading room was conducted at 3:45 p.m. No fires were found, and crews reported that areas of the building were cooling off.

Interior operations were concluded at 4:00 p.m. At 4:20 p.m., fire personnel, Watts Bar Nuclear Power Station Fire Brigade, Sequoyah Nuclear Power Station fire brigade, and fire investigation personnel from TVA met to determine scene security issues. One fire engine from Rhea County remained on-scene throughout the night in the event that additional fires might flare up.

Personnel preparing to enter the Control Building

Injuries and Fatalities

Five personnel working in the control building at the time of the fire suffered from smoke inhalation and two were transported to local hospitals. Upon the arrival of fire personnel, the building had been completely evacuated. No fire personnel were injured during fire suppression, overhaul, or investigation activities. Fire and investigation personnel were quick to point out that there could have been more injuries had the fire occurred later in the day. Had visitors been using the elevator at the time of the fire, they would have had no means of escape. Fire personnel also stated that the length of time it took to gain entry to the control building most likely helped p event njuries According to several firefigh ers, if they had been able to reach the terminal room on their first try, they would have been in very serious trouble from the heat and the amount of fire present

By the time fire suppression operations were concluded, over forty fire and medical personnel had responded to the scene. Unites from the Watts Bar Nuclear Power Station Fire Brigade, Wolf Creek Fire Department, Sequoyah Nuclear Power Station Fire Brigade, Evensville Fire Department, Pine Grove Fire Department Stations 1 and 2, Spring City Fire Department, Miegs County, and Rhea County Emergency Medical Services had worked together to ensure the successful extinguishment of the blaze.

Investigation into the origin and cause of the fire began the next morning, September 28.

THE INVESTIGATION

As soon as the control building was determined to be safe, a multi-agency investigation team lead by the TVA Police began their origin and cause determination. The team consisted of investigators from the Bureau of Alcohol, Tobacco and Firearms, Knox County Sheriff's Department TVA Office of the Inspector General, and TVA Police. Of top priority was ensuring that the fire was not the result of a deliberate act. Investigators prepared a preliminary report, which was released in November 2002. The report concluded that the fire was initially caused by a phase-ground-phase fault between one of the 480V cables running from the 480V main buses, and presented the results of a fire model run by the TVA Police.

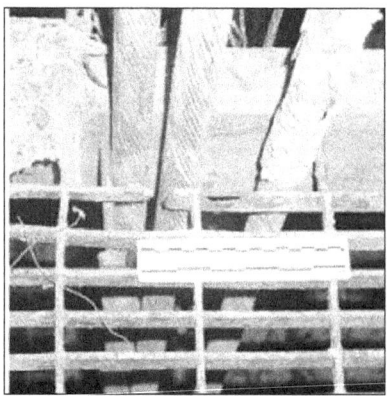

Point of fire origin

Based on damage to the cables and the steel decking, investigators concluded that prolonged rubbing of the steel decking against the butyl rubber insulation on the cable had resulted in a breaching of the cable insulation at the number five level (roughly 60 feet above ground) in the cable shaft. This resulted in an arc from the cables to the decking, and back to the cable The extreme temperature generated by this arc was enough to cause a small amount of the insulation to deteriorate into a flammable substance, thereby resulting in ignition. The large quantity of flammable materials in the shaft, coupled with a continuous chimney effect, provided the small fire with enough fuel to self-sustain and free burn. Within a matter of seconds, the fire started to spread rapidly. The high flammability of the insulation surrounding the cabling magnified the speed of propagation. Those cables which were covered with PVC released polyvinylchloride gas as they burned.

The construction of the cable shaft and the Control Building also helped the fire. The solid construction of the building trapped heat inside the structure, resulting in severe spalling within the cable spreading room, and causing temperatures on the terminal room and control room levels to skyrocket, even without visible flame. Based on the results of a fire model run by TVA Police, and using data from the United States Fire Administration, the National Institute of Standards and Technology, and the Nuclear Regulatory Commission, it was determined that temperatures within the control building reached or exceeded 1 200 degrees F during the fire. The heat generated in the shaft was so intense that steel decking 60 feet below the area of origin was severely warped, while water used during the first hours of attack instantly boiled on the floor of the structure or turned to steam.

Heat, smoke and flame generated by the initial fire traveled up the cable shaft, hitting their first obstacle in the cable spreading room. The sudden change in direction, lack of flammables, and compression of space could have smothered the fire. However, due to the advanced stage of the fire, it continued several feet down the cable channels into the switchyard and damaged cables as far as one hundred feet from the control building. In addition, the fire spread into the terminal level, igniting the fire load in the terminal room, with the result that firefighters initially considered it to be the room of origin and therefore the primary target of their attack. The smoke and the heat further penetrated the control room level, damaging everything on that level and causing further fires. Effects below the fire consisted of severe heat and smoke damage.

Damage Assessment

Although the fire occurred in a vertical shaft, there was a significant amount of damage to all parts of the hydroelectric plant. All cables running through the shaft were destroyed and severed as a result of the fire, resulting in a loss of communications fire annunciation systems, and most importantly, power transmission. Additionally, the ladders and steel decking in the vertical shaft were warped.

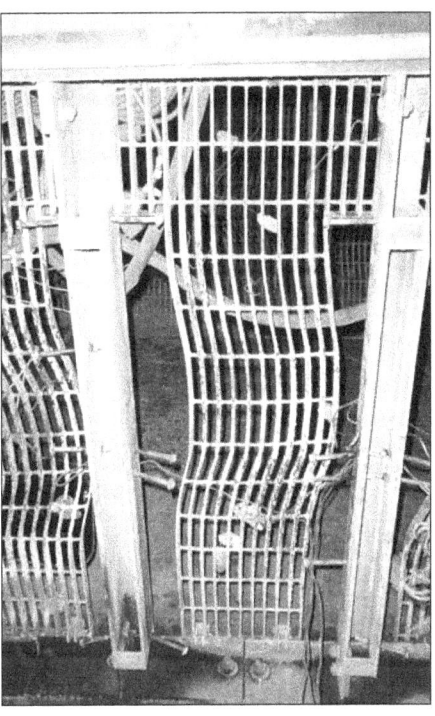

Warped decking below the area of origin

The cable spreading room above the shaft sustained significant spalling and smoke damage. The control building sustained significant damage, both from the effects of the fire and the effort associated with its extinguishment. Items within the building, including records, files, desks, breathing apparatus, and power control stations, were irrevocably damaged by the fire. In an effort to ventilate the building and reach the fire, firefighters forced doors and broke windows. Overhaul of the site resulted in further damage, as firefighters pulled ceilings and breached walls to ensure the fire was completely out. Water damage was also extensive, due to the significant volume of water necessary to cool the structure and extinguish the fire.

Control Building, approximately two hours into the fire

Due to the destruction of the power transmission cables, the generators had to be shut down. In order to prevent flooding upriver from the dam and to ensure that there was enough water flowing downstream to provide cooling for the Watts Bar Nuclear Power Station, the floodgates had to be opened. The closure of the generators and the opening of the floodgates have resulted in a significant loss of revenue for TVA. Estimates place the daily loss of revenue at up to $100,000.

KEY ISSUES

Building Structure

The Watts Bar Hydroelectric plant was constructed of concrete and steel, and located directly on the Tennessee River. The plant was essentially a bunker, with very few windows on the powerhouse floor, and small, sealed, windows in the lower levels of the control building. Even the large plate-glass windows located on the control room level of the control building were unable to be opened; their purpose was simply to provide a view of the dam for visitors. The building was comprised of various levels, connected by one staircase on the western side of the building The layout of the building, combined with the limited number of windows, made it a difficult structure to ventilate.

View of the control building from the South. Note the solid concrete construction and lack of windows.

The construction of the hydroelectric plant contributed to heat build-up and distribution. The concrete walls, floors, and ceilings acted as an oven, intensifying the heat and trapping it in the building. These conditions made firefighter entry and fire suppression extremely difficult. During fire suppression activities, firefighters reported that the water they were using to fight the fire was turning to steam or boiling upon contact with the floor. The level of heat in the building was further evident during the post-incident investigation. Investigators found heavy spalling in the cable spreading room.

The location of the building on a major body of water is significant for more reasons than power generation. Moving water generates a pressure differential within the local atmosphere. As a result, a slight breeze is produced This principle is commonly known as the Venturi Effect Firefighters are familiar with this principle through the practice of hydraulic ventilation – using a hose line to ventilate a building by spraying a fog pattern out a window, thereby creating a pressure differential and drawing air out of the room. The constant flow of water through the turbines in the powerhouse resulted in the Venturi effect in the Watts Bar Hydroelectric plant, with more damaging results. The control building was connected to the powerhouse by a vertical shaft, half of which carried cables, and half of which served as an elevator for visitors touring the plant The breeze generated by the constant flow of water through the turbines took the path of least resistance, moving from the powerhouse, through the cable shaft, and out the cable trays. This constant, ever-present breeze fed the fire in the cable shaft, resulting in rapid fire spread.

Within the hydroelectric plant complex, there are a limited number of egress routes from the control building On a normal day, in order to exit the building workers on the terminal room level are required to walk the length of the building through a variety of rooms and passages in order to reach the stairwell. They then have to climb a half-flight of stairs in order to reach the door. Visitors and workers on the control room level have easier access to an exit, although there is still only one on that level. An alternative exit for personnel on this level is to go downstairs to the exit used by personnel on the terminal level.

The day of the fire, firefighters would have had limited access to any personnel who might have become trapped on the terminal room level. While there were windows on the east side of the building there, they were small and very thick, making them a poor choice for entry. The only practical way for firefighters to enter this level was through the sole means of egress- -the southwest door. In the case of this fire, had any Watts Bar personnel become trapped on the terminal room level—a situation that nearly happened—they would have likely become casualties almost immediately.

Fire Protection Systems

The Watts Bar Hydroelectric Plant was constructed before the use of sprinklers were a standard part of building construction and was not originally designed to comply with what is the current NFPA 101 code. Even after the advent of sprinklering, there was little interest in installing sprinklers at this point. Conventional wisdom throughout the hydroelectric and the broader power generation community held that the fire risk to hydroelectric plants was minimal, based on their type of construction and proximity to water. As a result, retrofitting the building with fire sprinklers or suppression systems was not considered necessary. While the powerhouse floor was outfitted with a carbon dioxide (CO_2) dump system to suppress any fires that might occur in the generators, there were no fire suppression systems in the control building. Additionally, there were no detection or alarm systems in the control building. There was a detection system in the powerhouse, specifically in the hall storage room, generator cabinet, and governor cabinet. This system was connected to an annunciator panel in the control room, but the fire had effectively severed the connection rendering it inoperable.

Fire suppression services at Watts Bar Hydroelectric Plant are provided through long-standing mutual aid agreements with local volunteer fire companies in Rhea and Miegs Counties. While all personnel are familiar with these agreements, they are not set forth as formal, written policies. Rather, they are simply understood to be the way fire protection assistance is handled. In addition, the fire brigade from the Watts Bar Nuclear Station provides assistance when necessary.

Presumptions that hydroelectric power plants were essentially "fireproof" led to a decision in 1993 to reduce fire suppression personnel and training at the Watts Bar Hydroelectric Plant. For budgetary reasons, TVA decided to remove the on-site fire brigade from the plant and instead train plant personnel on incipient firefighting only.

TVA defined incipient fires as those that could be easily extinguished by one person, or a fire roughly the sizes of a small trash can fire. While this decision was reasonable based upon previous experience with hydroelectric plants, it left a gap in fire protection for the facility. Also missing was a formal plan for incorporating volunteer personnel into a fire response team that included or designated a command structure in the event of a fire. This lack of planning was evident on the day of the fire when personnel were initially unsure of who was responsible for incident command. While that uncertainty did not materially affect fire operations in this case, it is a point of concern. TVA, in their post-incident analysis, recognized this problem and is taking steps to develop and implement an incident management plan. They also have begun comprehensive program to bring all of their hydroelectric plants into compliance with the Life Safety Code and to set up fire suppression systems with the facilities.

Communications

The location of the Watts Bar Hydroelectric plant had a direct impact on fireground communications. The plant is located in a rural section of Rhea County in a very hilly and heavily wooded area. Although there are repeaters throughout Rhea and Miegs Counties, the radio signal in the immediate vicinity of the plant is still weak. The radios used by local response personnel were useful in communicating with other local personnel on the scene, but they did face some difficulty in clearly communicating with central dispatch.

Adding to the communications challenge, the fire brigade personnel from the Watts Bar Nuclear Power Station have radios with weak signals because material located within the nuclear power station is very sensitive to radio signals Their radios are useful for communicating directly with each other at the nuclear power station site, but they have a severely limited range. As a result, fire brigade personnel were unable to communicate with the nuclear power station from the hydroelectric plant site.

Some TVA fire brigade personnel have been con-ducting tests using cell phones with digital walkie-talkie capabilities as an alternative to using the radios. Several of these personnel were on-scene during the fire. However, the radio signal difficul-ties related to topography applied to the enhanced capabilities of cell phone as well. One member of the Watts Bar Fire Brigade stated that during the fire he found an approximately 10-square foot area in the north parking lot where he could use his phone. Whenever he needed to communicate with his phone, he had to go to this spot to get a signal.

Response personnel meeting face-to-face to discuss strategy

Another difficulty faced by personnel was the lack of interoperability of the portable radios. Local volunteer departments used one kind of radio, while responders from the Watts Bar Nuclear Power Station used another kind of radio that operated on a different frequency. Due to these differences, personnel were unable to communicate with each other by radio, so on-scene communications among response agencies were made face-to-face.. Personnel are aware that using radios is preferable, and local response agencies and the TVA are now planning to carry portable radios from the other response agency on their apparatus. In this way, Watts Bar Nuclear Power Station fire brigade personnel will have a Rhea County radio for use in an emergency, and Rhea County fire apparatus will have a Watts Bar Nuclear Power Station fire brigade radio.

There were also communications difficulties within the actual hydroelectric plant. The powerhouse and control building are linked via landline. When the fires started in the vertical cable shaft, com-munication lines between the control building and powerhouse were cut. The result was that person-nel in the powerhouse were completely unaware of the fire in the cable shaft and that people in the control building were being evacuated. Due to the constant breeze flowing through the powerhouse, all smoke from the fire was directed upwards and away from the powerhouse. The first indication that any powerhouse personnel had of a problem was when their power went out as a result of the fire There was no effective backup means of communication to warn workers of the fire and the pending need to leave the building. TVA recognized this problem in the after-action analysis of the fire, and is working to address it.

Risk

Personnel working within both the control building and the powerhouse faced challenges in exiting during the fire. As described earlier in this report the layout and design of the control building resulted in a limited means of egress for personnel in the control building. In the case of this fire, personnel had to pass above the fire to reach the exit. Had there been any significant delay, personnel in the building could have been seriously harmed trying to escape.

The design of the powerhouse also contributed to existing problems. The building is a large windowless structure, with heavy machinery and other dangers located throughout. It, too, has very limited means of egress, with exits to ground level on the west end and an exit to the top of the dam on the east end. The powerhouse has several levels, both above and below the waterline, with exits located on only one level. When power to the building was lost, workers were plunged into complete darkness. Personnel would have had to rely on their knowledge of the building and their sense of touch and hearing to exit the building. Luckily, a supervisor on

Powerhouse floor

the floor had a flashlight, and was able to find people and lead them safely from the building. TVA's review of life safety issues will likely include increased exit lighting per NFPA 101 7.8 1.

While personnel in both buildings had received training on evacuation, evacuation drills are not routinely held. Following the fire, TVA accident investigators determined that this was a problem that needed to be addressed, and they have recommended frequent training on and demonstrations of emergency and evacuation procedures.

Pre-planning

Local response personnel were aware that they would be the first responders to an event at the Watts Barr Hydroelectric Plant. Nevertheless, responders were not provided with plant diagrams, nor were there any plant pre-plans. Several of the local responders stated that they had only been in the plant on the visitor's tour, and had no idea what else was inside the building.

Control building diagrams were available at the site, stored in filing cabinets on the terminal room level. Due to the rapidity of fire spread and the need for quick evacuation, there was no time for personnel to grab the plans—they were lost in the fire. As a result, personnel working on-scene initially had to rely on employee's knowledge of the building to plan fire suppression activities and ventilation. Watts Bar Nuclear Power Station personnel requested that TVA fax copies of the plans to the site; these plans arrived several hours into the incident and were immediately utilized by on-scene personnel.

As a result of this fire, TVA and local response agencies are working together to develop comprehensive pre-plans for use in the event of another emergency at the hydroelectric plant. In addition, TVA is planning to provide local responders with a tour of the plant: the goal is to make this part of the annual training for responders.

Training

Prior to the fire at the hydroelectric plant, local responders, plant personnel, and the Watts Bar Nuclear Power Station fire brigade had not trained together. One of the firefighters from the fire brigade is a volunteer firefighter in Rhea County, but that was the extent of regular interaction among the response organizations.

In addition, personnel from the nuclear power station and local response agencies were unfamiliar with the various tactics necessary to fight a fire in the control building of the hydroelectric plant. Local responders have been trained to fight the types of fires they are most likely to encounter: fires in homes, brush or businesses Personnel from the nuclear power station have been trained to fight fires inside a nuclear power plant. While some of the tactics from each type of training can be used in fighting a fire in a hydroelectric plant control building, there are still special considerations and suppression tactics. Specifically, none of the responders were familiar with the special tactics needed to fight a fire involving an electrically charged object.

Fire in the breaker panels; looking from outside the Control Building, three hours into the fire.

TVA has developed tactics and methods for successfully fighting electrical fires using water. Some field personnel are trained on these tactics, and frequently use them to suppress fires. As a result of their post incident investigation, TVA recognized the necessity of training local response personnel who provide or support on-scene fire suppression in these specialized tactics. The challenge lies in finding the time and the funding to provide volunteer responders with this training. Volunteer firefighters stated that the best time for them to attend training would be during the weekend, when they are not at work. However, TVA stated that providing training on weekends would be a financially difficult option, due to the need to provide instructors with overtime pay TVA and local responders are working together on these issues, to ensure that responders receive the appropriate specialized training.

Local responders and TVA personnel also recognize the need to train together, in order to become familiar with the operating procedures and tactics used by each agency. Responders plan on drilling together more frequently and holding training exercises on TVA property.

Power Generation and Terminal

Upon arrival at the control building, TVA and local response personnel were uncertain as to whether the hydroelectric plant was still generating power. Due to the fact that the control building is the point where power from the hydroelectric plant is distributed to power lines, it was critical to determine if electricity was still flowing through the cables. Additionally, the building housed several backup batteries of varying voltage, which were supposed to provide power to the plant in the event of a sudden power loss. There was uncertainty during fire suppression operations as to whether these batteries had discharged their electricity, or if they were still charged.

Due to the uncertainty concerning the electrical situation, the decision was made by command personnel to treat all electrical apparatus within the control building as charged. This proved to be a prudent decision as it was discovered during the initial post-incident investigation that the batteries indeed were still charged.

Compliance with Life Safety Code

As the hydroelectric plant was finished in the early 1940's, the building was not originally designed to comply with the Life Safety Code (NFPA 101). In addition, there was no requirement that the building be retrofitted to meet the code. As a result, egress was severely limited, emergency lighting was not provided throughout the facility, and there was no mechanism for remotely activating a fire alarm.

Since the fire, TVA has recognized the importance of bringing all hydroelectric facilities into compliance with the Life Safety Code. While out of service, Watts Bar Hydroelectric Plant is undergoing renovations that will bring it into compliance. Additionally, the plant is undergoing comprehensive maintenance on all equipment.

FIRE ORIGIN AND SPREAD

During their initial investigation, TVA Police investigators worked with input from the United States Fire Administration, the National Institute of Standards and Technology, and the Nuclear Regulatory Commission to perform a fire engineering hazard analysis. According to the TVA Police record of investigation, this analysis "assesses the investigation of the fire's origin and cause, evaluates the performance of fixed fire protection systems, and identifies significant aspects of fire dynamics." Data generated by this analysis would then be used to help identify not only the fire origin and cause, but to determine specific hazards that might exist in similar plants.

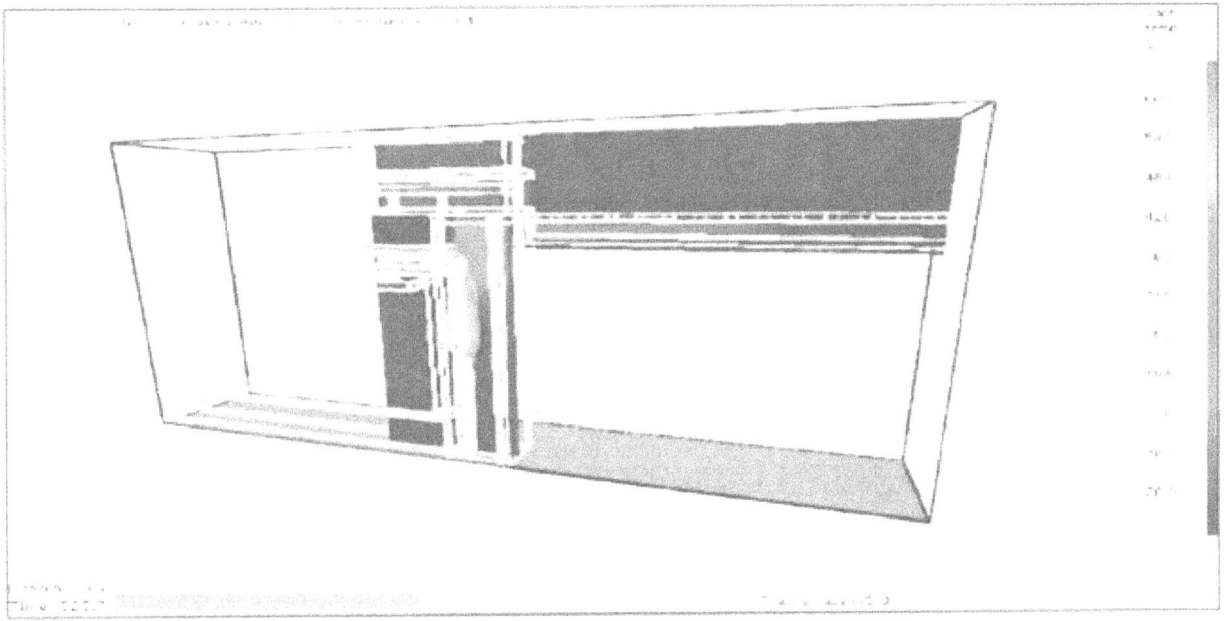

Three-dimensional computer model of WBH fire

TVA investigators utilized NIST's Fire Dynamics Simulator Version 2 9, inputting not only their own data but also information provided by USFA, NIST and NRC. The simulator modeled smoke and fire development and distribution over a period of time using a three-dimensional computer image to provide a second-by-second breakdown of the fire The results were telling. According to the record of investigation, "The model of the WBH (Watts Bar Hydroelectric) Plant after 527 second (8.78 minutes)…shows a fire plume extending up the cable shaft to the entrance to the ante room, which leads up a staircase to the cable spreading room."

Using information provided in the model, investigators were able to understand not only how the fire developed, but also why personnel working in the building had such short notice of the fire's existence. According to the report, smoke and heat from the fire hit the lower levels of the control building, "with a good portion of the heat and smoke being transferred out through the upper level cable shaft leading to the switchyard." Instead of smoke and heat spreading into the building, it was channeled down the path of least resistance, thereby, diverting any early warning that employees may have had. Smoke and heat penetrated the location of the employees only after the fire had built to such a fury that the path of greater resistance (the control building itself) could be penetrated.

The information this simulator generated was critical to developing a comprehensive understanding of fire development in a hydroelectric plant. It also served to illustrate the dangers associated with a fire in such a structure—namely, the rapidity of fire development, and the need for fire detection and protection systems.

LESSONS LEARNED

While the firefighting and suppression efforts at Watts Bar were successful, there were several lessons learned. These include the necessity of pre-planning, the importance of local firefighters and industrial firefighters conducting training exercises together, the need for adequate training of local response personnel in industrial firefighting, the need for effective primary and secondary communications methods, and the importance of compliance with fire codes.

Although plans for the Watts Bar Hydroelectric Plant existed, they were kept in the control building. There was no time for employees inside the building to collect the plans prior to exiting the building, so the plans were unreachable and unusable. Local fire departments were not provided with a copy of the plans, and so not only were they unfamiliar with the structure, but also did not have a visual guide to the design and layout of the building. Fortunately, several firefighters associated with the Watts Bar and Sequoyah Nuclear Power Stations and personnel from the hydroelectric plant were on-site during fire suppression operations, and were able to provide site and structural information verbally. It is critical that facilities relying upon local first responders ensure that these responders are provided with all necessary information to respond to and fight a fire at the facility. In addition to providing pre-plans to local response agencies, local responders should be given a tour of the facility, ideally on an annual or semi-annual basis.

Local first responders should become familiar with their counterparts at the facility, including any safety or fire suppression personnel who may work at the site. Not only should responders meet during site or facility tours, but they should also periodically drill on response procedures together. While the Rhea County emergency manager knew some of the personnel from Watts Bar, local responders were not familiar with their colleagues from the Watts Bar Nuclear Power Station fire brigade. Training together not only allows for personnel to meet each other, but it enables them to become familiar with any differences or similarities in response protocols and equipment that may exist between agencies.

In the case of the Watts Bar, an important difference exists between the firefighting training of local responders and that of personnel from the Watts Bar Nuclear Power Station. Local responders are trained in standard structural firefighting techniques, while personnel from Watts Bar are trained in industrial firefighting techniques. While these techniques all focus on fire suppression, they use different methods to achieve the goal. Moreover, fire personnel from TVA have received special training in techniques used to extinguish electrical fires, and are familiar with the hazards and limitations associated with fighting such fires. Traditionally, structural firefighters are trained to avoid fighting an electrical fire until the power source or the actual fire has been de-energized, or using a suppressant other than water. TVA employees, however, are trained to extinguish a wide variety of electrical fires using traditional hose lines. It would be beneficial for local responders to receive training, even alongside their industrial counterparts, in the special challenges and techniques of industrial firefighting. In this way, all response personnel would be familiar with the same techniques and objectives.

Another critical aspect of successful fire suppression is effective communications. Due to the geographical location of the hydroelectric plant, there were several "dead spots"—areas where radios cannot receive or send messages. Further complicating matters, the radios of local responders and TVA personnel were not compatible. Finally, the signal strength of the radios for the TVA Watts Bar Nuclear Power Station Fire Brigade was very weak. Although members of the nuclear power station fire brigade were within eyesight of the nuclear power station, personnel could not use their radios to reach it in order to keep supervisors there aware of their fireground operations.

On-scene TVA personnel were able to communicate using Nextel phones because of their digital walkie-talkie capabilities; however, these phones were only able to receive a signal in an area that was roughly ten square feet. Unit officers met face-to-face throughout the incident to communicate and make decisions. While effective in this instance, this means of communication is not always practical or reliable. If financially feasible, local response units should be provided with a radio from the industrial site, thereby ensuring that a means for radio communication exists. Additionally, the industrial site should be provided with a radio from the local response agency. These radios could be carried in response units, and used exclusively for communication during an incident or drill.

One of the most important lessons learned from this fire is the importance of compliance with the Life Safety Code. Although the Watts Bar Hydroelectric plant was grandfathered in due to its age, non-compliance with the code could have had fatal results in this fire. As mentioned before, personnel on the powerhouse floor did not know there was a fire burning in the cable shaft. There was also no means for warning them, as communication between the control building and powerhouse had been cut by to the fire. There are very few means of egress from the powerhouse, and several places where an employee could be trapped and seriously injured or killed. Worse, there was no emergency lighting, and employees were plunged into total darkness when the power failed. Only the quick thinking of a floor supervisor ensured the safe exit of employees. As a result of this fire, TVA has realized the importance of the life safety code, and plans to bring all of its hydroelectric plants into compliance.